Anam Khalid

Spatial and Temporal Distribution of Dengue

A Case Study of Lahore

Anchor Academic
Publishing

Khalid, Anam: Spatial and Temporal Distribution of Dengue. A Case Study of Lahore, Hamburg, Anchor Academic Publishing 2016

Buch-ISBN: 978-3-96067-053-7
PDF-eBook-ISBN: 978-3-96067-553-2
Druck/Herstellung: Anchor Academic Publishing, Hamburg, 2016

Bibliografische Information der Deutschen Nationalbibliothek:
Die Deutsche Nationalbibliothek verzeichnet diese Publikation in der Deutschen Nationalbibliografie; detaillierte bibliografische Daten sind im Internet über http://dnb.d-nb.de abrufbar.

Bibliographical Information of the German National Library:
The German National Library lists this publication in the German National Bibliography. Detailed bibliographic data can be found at: http://dnb.d-nb.de

All rights reserved. This publication may not be reproduced, stored in a retrieval system or transmitted, in any form or by any means, electronic, mechanical, photocopying, recording or otherwise, without the prior permission of the publishers.

Das Werk einschließlich aller seiner Teile ist urheberrechtlich geschützt. Jede Verwertung außerhalb der Grenzen des Urheberrechtsgesetzes ist ohne Zustimmung des Verlages unzulässig und strafbar. Dies gilt insbesondere für Vervielfältigungen, Übersetzungen, Mikroverfilmungen und die Einspeicherung und Bearbeitung in elektronischen Systemen.

Die Wiedergabe von Gebrauchsnamen, Handelsnamen, Warenbezeichnungen usw. in diesem Werk berechtigt auch ohne besondere Kennzeichnung nicht zu der Annahme, dass solche Namen im Sinne der Warenzeichen- und Markenschutz-Gesetzgebung als frei zu betrachten wären und daher von jedermann benutzt werden dürften.

Die Informationen in diesem Werk wurden mit Sorgfalt erarbeitet. Dennoch können Fehler nicht vollständig ausgeschlossen werden und die Diplomica Verlag GmbH, die Autoren oder Übersetzer übernehmen keine juristische Verantwortung oder irgendeine Haftung für evtl. verbliebene fehlerhafte Angaben und deren Folgen.

Alle Rechte vorbehalten

© Anchor Academic Publishing, Imprint der Diplomica Verlag GmbH
Hermannstal 119k, 22119 Hamburg
http://www.diplomica-verlag.de, Hamburg 2016
Printed in Germany

Abstract

This study was conducted to evaluate the dengue outbreaks pattern in spatial and temporal context and to identify the meteorological constraints behind the widespread so that a complete picture of the scenario may be developed. The Lahore District was selected as Study area which was the majorly affected by mosquito on a greater extent. Lahore is the second largest city of Pakistan with respect to population due to its rapidly growing urbanization. The environmental factors affecting the spread of the disease have been identified and then mapped in a GIS based environment using all the spatial and tabular data obtained from different sources. The factors affecting dengue spread were found to be Land Surface Temperature (LST), Land cover/Land use, Normalized Difference Vegetation Index (NDVI), Temperature, Rainfall, Population Density.

The results depicted some particular trends. Areas with high population density were found to be more infected as more people got affected Some areas with comparatively low population density were also found t be infected from the mosquito for the reason: they were actually the high income areas with huge houses and lower number of residents. They mostly contain lawns and swimming pools which are the most active breeding sites of the mosquitoes.

Another major factor incorporated in the study is, the temporal factor. There was a drastic change in terms of number of patients was found in the years 2011 and 2012. To investigate the reasons, all the meteorological and clinical data of both years has been compared. It concluded that through the pre-planned launch of the control activities, both from the City Government and the Civilian side, resulted in such a controlled number of the victims. Going in the same direction with the same spirit may nullify the virus.

Acknowledgements

Deep regards to The Urban Sector Policy and Management Unit, P & D Department, Govt. of the Punjab for providing all the spatial data required to make this research possible.

Deep gratitude to the Pakistan Meteorological Department, Lahore for facilitating with the climatic data regarding rainfall and temperature.

I am especially grateful to my supervisor, Dr. Syeda Aadila Batool for her most valuable guidance and advices throughout the research. I would not be able to complete my research with her diligent concern and continuous encouragement.

Lastly, I want to thank my family and my friends for their support and being patient when I was low during the toughest part of the research.

Contents

Abstract
Acknowledgements

List of Figures ... vi
List of Tables ... viii
List of Graphs ... ix

Chapter One — Introduction

1.1 Background ... 1
1.2 Need of Research ... 2
1.3 Scope of Research ... 3
1.4 Problem Statement ... 5
1.5 Objectives and Work Flow ... 5
1.6 Conceptual Framework ... 7

Chapter Two — Review of Literature

2.1 What is Dengue? ... 8
    2.1.1 Types of Dengue ... 8
    2.1.2 Symptoms of Dengue Fever ... 8
2.2 Related Literature ... 9
    2.2.1 Spatial Risk Model ... 10
    2.2.2 Spatial correlation with socioeconomic and environmental variables ... 11
    2.2.3 Analysis of Dengue Spreading Dynamics ... 12
    2.2.4 Spatio-Temporal Mapping of Dengue Spread Pattern ... 14
    2.2.5 Geostatistical Modeling ... 15
    2.2.6 Modeling of Seroprevalance of Dengue and its Vector Density ... 16
    2.2.7 Mapping of Temporal Risk Characteristics ... 17

Chapter Three — Study Area

3.1 Profile of Study Area ... 18

3.2	Geographic Location of Study Area	19
3.3	Land Cover of Study Area	20
3.4	Hydrological Layer of Lahore City	20
3.5	Climatic Conditions	21
	3.5.1 Temperature	21
	3.5.2 Rainfall	22
3.6	Suitable Breeding Sites for Dengue	23

Chapter Four   Materials and Methods

4.1	Data Collection	24
	4.1.1 Epidemiological Data	24
	4.1.2 Population Density	25
	4.1.3 Identification of Environmental Parameters	26
	4.1.3.1 Precipitation	27
	4.1.4 Data from Landsat V	27
4.2	Methodology	27
	4.2.1 Land cover/ Land use	27
	4.2.2 Land Surface Temperature (LST)	28
	4.2.3 Normalized Difference Vegetation Index (NDVI)	30
	4.2.4 Spatial Units to analyze and map Dengue Incidence data	31
	4.2.5 Temporal Units	32
4.3	Data Analysis	32
4.4	Identification of risk areas	33

Chapter Five   Results and Discussions

5.1	An Overview	34
5.2	Spatial Pattern of Dengue Occurrences	34
	5.2.1 Land Surface Temperature and Dengue Incidences	34
	5.2.2 Land Cover Map against Dengue Occurrences	35
	5.2.3 Normalized Difference Vegetation Index	37
	5.2.4 Population Density	38

5.3	Identification of High Risk Areas Using overlay Analysis for 2011	40
5.4	Temporal Analysis between 2011 and 2012	42

Chapter Six   Conclusions and Recommendations

6.1	Conclusions	44
6.2	Limitations	45
6.3	Future Recommendations	45

References   47

List of Figures

Chapter One   **Introduction**

Figure 1.1	Conceptual Framework of the Research	7

Chapter Three   **Study Area**

Figure 3.1	Geographic Location of Study Area in the Country	18
Figure 3.2	Map of Lahore District with its UC boundaries	19
Figure 3.3	Map of Hydrological Layers of Lahore District	21

Chapter Four   **Materials and Methods**

Figure 4.1	UC Wise population density of Lahore District - 2011	26
Figure 4.2	Land cover Classification done in ERDAS Imagine and mapped in ArcGIS 9.3	28
Figure 4.3	Land Surface Temperature for September, 2011	29
Figure 4.4	Land Surface Temperature for September, 2012	30
Figure 4.5	Normalized Difference Vegetation Index (NDVI)	31
Figure 4.6	Map showing the most critical areas for dengue spread	33

Chapter Five   **Results and Discussions**

Figure 5.1	LST of 2011 with No. of patients in all UCs of Lahore	35
Figure 5.2	Land Cover Categorization of Landsat V image for Sept, 2011	37
Figure 5.3	NDVI in Comparison with the number of patients for Sept, 2011	38
Figure 5.4	Population Density compared with the population density	39

Figure 5.5	Accumulative map for spread pattern of dengue with all its causal parameters	40
Figure 5.6	Accumulative map of dengue Spatial spread pattern with all the parameters	41
Figure 5.7	Dengue Spread pattern for the year 2011	42
Figure 5.8	Dengue Spread pattern for the year 2012	43

List of Tables

Chapter Three **Study Area**

Table 3.1: Land Cover Classification with its percentage   20

Chapter Four **Materials and Methods**

Figure 4.1 Monthly no. of reported patients of Dengue   25

List of Graphs

Chapter Three   **Study Area**

Graph 3.1   Temperature curve for the year 2011   22

Graph 3.2   Comparison of rainfall in mm of the years 2011 & 2012   23

Chapter Four   **Materials and Methods**

Graph 4.1   Graphical representation of patient count for 2011   24

Chapter 1
Introduction

1.1 Background

Dengue Fever (DF) and Dengue Haemorrhagic Fever (DHF) are important public health concerns in the tropic and sub-tropic areas all over the world, during past 50 years. Rapid urbanization, increasing population movement, Global Warming and Drastic Climatic changes that contribute to the proliferation of man-made habitats of the mosquito are the prominent factors for the increasing numbers of Dengue incidences. The DF and DHF occur over 100 countries throughout the Americas, Southern Europe, North Africa, Mediterranean, Asia and Pacific regions.

It is the most common vector-borne viral disease of humans worldwide with an estimate that 50 million infections occur annually, with 500,000 cases of DHF which is an even more dangerous form of Dengue infection and at least 12,000 deaths. (Umor S et al, 2002). As far as the most affected areas are concerned, prevalence of Dengue is highest in Tropical areas of Asia and America, with 50 to 100 million estimated cases of DF and 250,000 to 500,000 cases of DHF occurring annually worldwide as explosive outbreaks in urban areas. (Chinnock, 2009)

This Tropical and Sub-Tropical disease is most prominent in urban and semi-urban areas. DHF, a potentially lethal complication, was first recognized in 1950s in the Philippines and Thailand, but today DHF affects most Asian countries predominantly in the developing countries where environmental condition is quite poor and health facilities are insufficient. DHF is now becoming a leading cause of hospitalization and death, majority of them are children. (Bhandari K et al).

Dengue has been the talk of the town during past almost five years in Pakistan. The cause of its fame is its disastrous mortality rate, its rapid spread and the after effects if the patient survives. It is a viral disease rather a tropical viral disease that is caused by an infectious mosquito named Aedes Aegypti that directly attacks the immune system of the infected person. Dengue has been a part of Pakistan since 1982 but during last two years it took form of a monster that horrifies everyone with its wide spread across the country. Lack of efficient prevention planning techniques, insufficient health facilities, poor sewerage systems, bad quality sprays, and hazardous environment supported this monster making it even more powerful. Moreover, the most alarming fact about Dengue is about its breeding sites. It is a clean water mosquito that breeds well over clean stagnant water.

The DF/DHF is a Public Health problem in Pakistan where the number of such reported cases has dramatically increased during the last ten years. Remote Sensing and Geographic Information System (GIS) technologies have been used in this study to link and update information on the environment, weather conditions, reported number of Dengue cases and other prominent factors. These technologies have been widely used in Public Health sector for managing and monitoring the problem. Remote Sensing data is utilized to manage the problem by incorporating population data and environmental factors such as changes in land use/land cover, land surface temperatures, rainfall etc.

1.2 Need of Research

Dengue Infection (Anti-D3) was first documented in 1982 from Punjab (Central Province of Pakistan) in 12 patients out of a blood sample of 174 blood samples of different people collected in 1968 and 1978. The first reported outbreak of Dengue Haemorrhagic fever in Pakistan was in 1994. According to a recent study, WHO has classified Dengue Infection in Pakistan into Dengue Fever, Dengue Haemorrhagic Fever (DHF) and Dengue Shock Syndrome (DSS). Out of 225 cases (particularly taken for research purposes) 137 (61%) were found to have

Introduction   Chapter 1

Dengue Fever, 81 (36%) were having Dengue Haemorrhagic Fever (DHF) and 7 (3%) were having Dengue Shock Syndrome (DSS). During year 2000 to 2004, 73% of Dengue Infected patients had Dengue Fever, 24% had DHF and 2.4% had DSS whereas after 2005, the ratio of DHF was increased by 39% while 58% got Dengue Fever and 3% got DSS.

Punjab got the highest outbreak of Dengue and more precisely, Lahore got the maximum number of patients/deaths reported during last 2 to 3 years. The actual figures are far greater than the ones we got from the clinical data which included the number of people who got their treatment at home by self-medication. Lahore is a very large district having nine towns which are further subdivided into 150 Union Councils each with slightly different environment in urban perspective. Among them, some areas are having wealthier neighborhoods having more greenery and hence a better and healthier environment but still was the hot spots for dengue outbreak which causes confusion for health Authorities and simulated a lot of researchers to find the reasons.

Dengue surveillance and control in large urban areas with high levels of dengue transmission are the most important challenges. Clinical surveillance is impaired by the high proportion of asymptomatic infections, and mosquito surveillance is very time consuming and resource consuming task. Moreover, despite the theoretical and numeric relationship between vector abundance and risk and potential of transmission is still poorly known. And, due to unavailability of vaccines or any particular medication, this quantitative assessment is quite necessary to know. (Chinnock, 2009)

1.3 Scope of Research

Dengue fever over the last 5 years has become one of the most serious diseases in Pakistan. Dengue is an arboviral disease transmitted by a mosquito called Aegypti having four serotypes, i.e. DEN-1, DEN-2, DEN-3, and DEN-4. It is a pre-

domesticated mosquito that prefers to breed in clean water sites such as artificial containers, swimming pools, ponds etc. Lahore has been the major target of this perilous virus where the count of reported cases was maximum throughout Pakistan. Considering the fact that there is no vaccine available for this severe viral disease, it is very important to map dengue outbreak dynamics with all the environmental parameters for the prediction of high risk areas and conditions under which this virus persists in order to take precautionary measures.

Secondly, mapping the current situation of patient count along with their location may also help to properly plan the medical facilities. It has seen that, in areas with a wealthy population, the access for medical facilities is not a big deal and hence it directly influences the fatality rate, no matter the number of infected people is lesser or greater. Pakistan being a developing country, lack in financial resources and there for it is the need of the hour to launch an effective planning program which should be efficient enough so that wastage of finance, resources and efforts could be avoided. Mapping of accurate patient count along with their access to medication would definitely help in achieving this goal.

Geographic Information System (GIS) and Remote Sensing can efficiently be used for disease mapping to highlight the high risk areas for pre-planning against spread of that particular epidemic to avoid a large number of fatalities. GIS has made it easy to analyze the clusters of reported cases with their geographic location and the environmental parameters of the location where the major outbreak took place and the prediction of other high risk areas with temporal dynamics of different epidemics, which further incorporated with different types of data may help to launch a planned medical.

Advances in Remote Sensing technology and GIS have enabled us to get crucial information and their modeling on Dengue transmission. Remote Sensing provides up to date information for Land use and Land change, Vegetation, Surface Temperatures, and Water Bodies etc. Whereas GIS enables us to model all the available information for a better understanding and illustration of all phenomenon taking part in the spread of the disease as well as prediction and planning for the prevention in future outbreak areas.

1.4 Problem Statement

A lot of our Administrative Authorities are working day and night to overcome this disastrous outbreak of Dengue in as planned manner as possible. The need to figure out the actual reasons why Dengue appeared on this part of land is as important as the need of provision of effective health facilities. The reason why some places are best breeding sites of Dengue, which kind of environment enhances its breeding, its cycle (i.e. the peak months of Dengue outbreak) and the pattern on spread is very important to know. Are the expected parameters really affecting the breeding of Dengue or is it just a myth? All these problems are yet to be known and are quite significant so that, with the passing years, the loss of precious lives could be avoided.

1.5 Objectives and Work Flow

About 2500 million people are reported to live in the regions with the estimated risk greater than 50% for Dengue transmission, one of the world's most widespread vector-borne diseases. An Empirical Model Analysis also shows that, if climate stays mild, about 3.5 billion people, 35% of the population will be at risk of Dengue transmission in 2085, and about 5 to 6 billion people (50 to 60% of the projected global population) is the change worsens. (Hales et al., 2002). Unavailability of vaccines has made the situation even more dreadful. The mortality rate in developing countries is more alarming as compared to developed countries due to poor health facilities and certain environmental factors. The only cure for this vector borne disease is no cure and hence prevention has become null and void.

A quick launch of control activities has already been done but prediction of high potential risk areas is still a major issue due to the great diversity in the epidemiological pattern of the DHF which makes DF difficult to predict.

Two main patterns may describe the fluctuations of in occurrence of DHF. The cyclic pattern refers to the seasonal variations of transmission. The viral disaster reaches a peak during the hot and rainy season (conditions being suitable for breeding of Dengue Mosquito). The end of the rainy season results in a return to a lower level of transmission. This phenomenon is repeated every year and characterizes the regional mode of transmission. This Cyclic Pattern of Dengue or more precisely called the Dengue Wave has been identified in this study using the clinical data. (Barbazan. P et al. 2000)

Secondly, another range of temporal changes has been identified, taking two years i.e. 2011 and 2012 as sample for our study. The clinical as well as meteorological data has been compared between these two years to figure out the causes behind the change between these two consecutive years. In 2012, a drastic decrease in number of patients has been analyzed which was quite surprising. The later sections include the changing patterns and the caused behind them. The temporal analysis was aimed to map the changing trends and to make the results of control and prevention so that it may become a sample for other affected areas

Introduction   Chapter 1

1.6 Conceptual Framework

Developing GIS Database

Demographic factors / Clinical Data:
- Population Density of the area
- Age
- Gender
- Residential address (to get the Spatial location of the patient)

Environmental Factors:
- Land Surface Temperature
- Precipitation Data
- Air Temperature (in case of Unavailability of LST)
- Normalized Difference Vegetative Index (NDVI)

Spatial Data
- Land Cover Classification
- Spatial locations of the reported Patients

Data Processing and Results

- All the acquired data from different sources has been brought into GIS Environment.
- A part of the metrological data and Spatial data has been extracted from high resolution satellite imagery
- All the acquired data has been mapped and analyzed with the help of different GIS techniques in order to achieve our desired goals

Figure 1.1: Conceptual Framework of the Research

Chapter 2
Review of Literature

2.1 What is Dengue?

Dengue is an infectious viral disease that is transferred by a bite of the infected female mosquito named Aedes Mosquito. It is also called as break-bone fever because in some cases it causes severe muscular and joint pain in the patient. (WHO report, 2012)

2.1.1 Types of Dengue

There are four major types of Dengue found all over the world which are termed as:

- DEN 1
- DEN 2
- DEN 3
- DEN 4

Each type is severe than its predecessor. Unfortunately, there is no awareness against these serotypes of dengue and all the infected people are treated as the infectants of the same virus. (WHO report, 2012)

2.1.2 Symptoms of Dengue Fever

It is a flu- like illness that may infect people of any age group and any gender. Its symptoms include:
- Mild to quite incapacitating high fever
- Severe headache
- Pain Behind the eyes
- Pain in muscles
- Pain in joints
- Rashes (normally in hands and feet)

The symptoms appear in 3 to 14 days after the bite of the mosquito. The condition of the patient depends upon the serotype of the virus, the patient is infected with. There is no specific medication or vaccine available for the cure but it is recommended to maintain the hydration level in case of dengue.

In case of sever dengue, plasma leakage may cause organ failures or in some cases, lead to death. Early clinical diagnosis and careful medical treatment may lower the risk to the lives.

2.2 Related Literature

Dengue Fever along with all its Serotypes are the most dreaded mosquito-borne viral disease since last almost 3 years. The cases are reported throughout Pakistan. Dengue hit Lahore with a greater intensity as compared to the other districts of Pakistan where the number of cases have dramatically increased within no time.

The same situation has occurred in many countries all over the world alarming the Govt. to plan out effective health safety activities to avoid the causalities to its minimum. Remote Sensing and GIS technologies have been widely used for this purpose to link the updated information of the patients with the other important factors that include environmental as well as socio- economic parameters. Identification of the high risk areas based on all the socio- economic and demographic patters not only present the current situation but also demonstrates where the health facilities are actually needed and eventually help avoiding the clustering of health facilities over the same area.

A number of studies have been done all over the world to identify the high risk areas based on different socio-economic and environmental parameters as per the objectives of the study. Some of them have been used as a model for our study which is as follows:

2.2.1 Spatial Risk Model

A case study was in Taiwan to examine the Spatial and Temporal patterns in Kaohsiung in 2009 by Hung Wen T et al. The basic purpose of the research was to identify the high risk areas where the outbreak originate and went to its maximum level for early prevention and control. The combination of GIS and statistical analysis gave considerable results in terms of identification of hot spots for dengue prevalence. It included the analysis based on the patient data recorded in the year 2002.

Instead of relying annual incidences to map the epidemic, they used a spatial risk model integrating a number of epidemiological characteristics of that outbreak and identified both the temporal and spatial risk factors related to the high potential risk areas.

The major epidemic period was identified during the year 2001 to 2003 to be the year 2002 (since January 2002 to December 2002). After identifying the time frame, the risk patterns were identified using Dengue Statistical data of Taiwan mapped using ArcGIS 9.1. The study used the smallest Local Governing Unit in Taiwan, named "Li" and a week (7-day week) as the temporal unit. They mapped a total of 423 Li units within 52 weeks of 2002. Duration of epidemic and intensity of transmission of epidemic was also mapped which was taken as, the total number of dengue waves during the entire defined time period. An amalgam of all these parameters yielded the temporal map of dengue wave throughout the study area.

Population risk levels were figured out by using spatial auto correlation among the neighborhood of the infected areas. The risk levels range from extreme risk level to low risk level based on the quality of the targeted area and its neighborhood. Using both temporal and spatial data evaluated in two different steps were then combined to form a cumulative Spatial and Temporal map depicting an overall picture of the current condition.

A relation between the size (area) of the study area was also examined along with the population of that area. It is a very positive approach as greater number of patients may refer to greater area and consequently a larger population and vice versa. It may falsify the perception for a particular area being apparently highest in the number of patients. To summarize, this study used the GIS in combination with the simple statistical analysis to analyze dengue risk areas, its pattern of transmission in relation with the population.

It is an excellent study demonstrating the risk areas of the particular study area providing visual model for risk management and planning against dengue epidemic.

2.2.2 Spatial correlation with socioeconomic and environmental variables

Mondinj A et al. in 2008 mapped the dengue risk areas but with a different range of parameters. The research included Census data being the basic parameter for identification of the potential areas. The basic aim was to develop to map a model of spatio-temporal risk of dengue cases and identify the threatened areas. The data for dengue patients was divided into two sections/ periods, i.e. from September 1994 to August 1995 and from September 2000 to August 2001 being the years having greatest number of patients. The analysis was based on respective annual incidences of dengue. Different demographic patterns such as, income levels, education levels, proportion of illiterate people, density of homes per mile etc. the environmental variables include waste collection, water supply and sewage coverage. Thematic maps are made in GIS environment to show the spatial distribution of dengue incidences in different Census tracts. The maximum threatened zones of the city were identified with the help of these maps.

The study mainly focuses over the socio-economic factors of the region. These socio-economic factors include single-storey homes, number of residents per home, percentage of homes without domestic water supply, low income homes and number of homes without waste water collection. Spatial analysis was performed over the maps to depict the role of socio-economic, demographic and environmental variables over the whole region.

Talking about the results, it is evident from the study that a greater number of cases were reported in the low socio-economic areas having incredibly low income. The proportion of one-storey homes was taken to identify the construction pattern over the region and to analyze the relation between the construction patterns with the number of dengue incidences. Well planned areas generally refer to posh and wealthy areas of a city. And it is a common practice that these areas are normally entertained by better health facilities resulting in reduced disease occurrence rate and death rate.

Incorporating construction patterns give excellent ideas to highlight which areas are at high risk and which ones are yet to at this risk level. Their results showed an inverse correlation between the income and the number of dengue incidences. Low income areas have higher number of dengue patients and vice versa. The reason normally is lack of health facilities and awareness regarding precautionary measures. Whereas, a direct relation was observed was observed between one-storey homes and the number of prevalence. This study aids the Authorities to make public policies aiming to improve socioeconomic indicators that would impact on dengue transmission.

2.2.3 Analysis of Dengue Spreading Dynamics

Another study was carried out to analyze Space-time analysis of the dengue spreading dynamics, laid for Northern Argentina in 2004. This research has an approach towards the combination of biotechnology and GIS. The data collection and initial analysis over the data is more towards biological side whereas, the mapping was done in GIS environment. Laboratory testing was performed on the serum samples for a certain number of cases. These serum samples (on being positive) were then further classifies into different types based upon the laboratory results.

Then there comes a stage of mapping dengue incidences based of different parameters. The first parameter mapped, is the age of the patients. It was observed that the highest incidence occurred in the people of age group 15 to 29 years while a lower number of incidences were found in the people of the age lower than 14 years and greater than 45 years. The temporal maps were generated taking week as the smallest unit. The hot spots over the whole study area were mapped and it was noticed that after the launch of health services notification, a faster expansion of the disease occurred in the villages as compared to those in urban areas. The reason behind this scenario is, villages normally have insufficient health facilities, so, as the health facilities were provided, people approached the medical camps to cure themselves and hence the cases were reported. It doesn't mean that dengue spread expanded after the provision of health facilities; rather it was due to reporting.

The dengue cyclic pattern has also been observed by mapping 487 confirmed dengue cases in a wave pattern. Two important peaks were found by the mapping, one at 3 days duration and the other at 12 to 15 days duration. It means that the outbreak persisted for the mentioned period of time.

The third dimension of the research is the predictive model based on the facts and figures gathered from different sources. This predictive model was made in ENVI software using Linear Pearson Correlation Coefficient. A simulated incidence was formed by taking the existing figures into account. This showed 80% correlation and provided a predictive mop for planning against the disease even before its outbreak began once again.

It figured out the reason behind a certain age group people being the major victim of dengue. The reason was, mobility can be seen among the people of age group between 15 to 45 years (being young). Hence this mobility in young adults bring them into in contact with the hit spots of dengue presence. Furthermore it has also

Review of Literature   Chapter 2

been found the risk of dengue is greater at homes because of endophilic behavior of the mosquito.

The space-time mapping showed three different patterns of clusters. The reason behind the first cluster of one day temporal range and 100 m spatial range may because the single infected mosquito bit people to complete its feed. The second cluster may be due to multiple female mosquitoes and the extension of the period. The mosquito survival days may be extended with an EIP (Extended Incubation Period). The reason behind the third cluster was said to be similar to this one.

In a nut shell, this study highlights the benefits of using GIS and remote Sensing technologies in efficient disease Surveillance System planning. However, using Remote Sensing for dynamic spatial modeling is linked to macro factors (large scale factors), that is able to explain some variables of outbreak behavior but not all and these kind of models cannot be used alone. Rather they require collaborative studies with the epidemiologists, ecologists and health professionals. This kind of study may be generalized to map other outbreak events as per feasibility.

The methodology and tools developed for this study has the potential to help in describing time cycles of the disease and simulating the outbreak speed in certain areas where no field data is available. This study can help decision makers to improve health system responses and prevention measures in relation with the vector control.

2.2.4 Spatio-Temporal Mapping of Dengue Spread Pattern

This research was held to study the dengue fever outbreak pattern in Lrqcoubo, French Guiana by Tran A et al. For this purpose, the location of all the patients' homes along with the date of these reported cases were observed. GIS was used to the information related to the patients. Spatiotemporal clustering was detected and then analysis of relative risk variations was then observed.

This study is again a combination of biotechnology and GIS and Remote Sensing. Initially, serological testing of the patients was performed to get the serotype of the disease. Then the Spatial-Temporal patterns Analysis was performed over all the collected data. The data included no. of patients with their spatial locations, the land use of the study area, average distance between each dwelling unit of the study area. The analysis shows a relationship between areas at high risk and the temporal risk over there.

The study on space time patterning enabled to map the RR (Relative Risk) within a particular space-time window. The RR map helped to extract boundaries of the areas having highest risk levels.

The study proves the face that Dengue virus spread using GIS and Space-time statistics allows epidemiologists to define and point out risk areas which is necessary for implementation of an efficient surveillance strategy. The maps formed for the study were on weekly basis with highlighted risk areas which were presented to different stake holders to improve prevention measures. Secondly, this information was linked to the relevant environmental factors to establish a model of the epidemic.

The findings of the research demonstrate the relevance and potential of the use of GIS and statistical analysis for elaborating a dengue fever surveillance strategy.

2.2.5 Geostatistical Modeling

Seng et al. Modeled, analyzed and mapped dengue epidemiology to understand the correlation between dengue fever occurrences, population distribution and meteorological factors. By combining GIS with geostatistical analysis, the spatial variation of dengue incidence can be mapped. This analysis showed a strong positive relation between dengue prevalence and population distribution in the study area. This demonstrated a higher number of dengue patients in densely populated urban area as compared to low dense areas.

The study also focuses on the environmental factors suitable for the breeding of dengue mosquito. It identified that only 100 to 14 days of rainfall is sufficient to support the breeding of the mosquito. It extracted the correlation between population density and the extent and spread of dengue fever. Keeping in consideration the topology type, rural and urban conditions, Solid Waste Management System, Forest covered area and agricultural land; it has deduced some valuable results. These findings include:

- The areas with a larger population are supposed to have a high dengue incidence rate.
- Two areas with the same population density and environmental constraints were found to have different level of dengue incidence. The lower incident area was due to control and prevention activities.
- Urbanization and poorly managed solid waste has found to be a major cause of dengue transmission.
- The breeding sites were found to be the developing areas of the city.

2.2.6 Modeling of Seroprevalance of Dengue and its Vector Density

Brazil has also been victimized by dengue. A severe dengue outbreak occurred in 2008. This research was conducted in order to map the pattern of dengue spread and correlate it with socioeconomic profiles/ parameters. For this purpose, they have divided the whole study areas into three major categories:

- Central Urban Area
- Sub Urban Area
- Slums

It was found that the highest seroprevalance occurred at slums. The most noticeable fact is, the mosquito density was lesser in slums but found to have a greater number infected people there. The reason being quite simple, that the condition of the household was quite suitable for the spread/ interaction between the victim and the mosquito. Secondly, another trend was observed, that the high risk areas were those with a higher human movement. This depicts a very important fact that the humans may also be responsible for the spatial widespread of the dengue throughout the area.

2.2.7 Mapping of Temporal Risk Characteristics

This study also focuses on the spatial and temporal risk factors to model against the spread of this epidemic disease. A spatial vs. temporal pattern relationship of dengue spread has been found out with the help of adequate amount of detailed statistical data. It concluded with two types of trends:

- Areas with a stronger intensity for short duration
- Areas with varying intensity but longer duration.

On the basis of the identification of these areas, the spatial cluster mapping of the dengue incidences was performed over the whole district. A block with extremely high rate of dengue prevalence may affect its neighboring areas. On the contrary, if some low affected area is surrounded by high affected areas, it also has the potential to adopt the same trend as its neighboring areas have. Once these areas are identified, they will lead to the assessment of the type of prevention activities in the affected area. Identification of the potential areas may also help to avoid the high mortality rate as, applying the control and prevention measures may become effective and save precious lives.

Chapter 3
Study Area

3.1 Profile of Study Area

A descriptive study was conducted in one of the Eastern District of the Punjab province i.e. Lahore. The reason behind the selection of this district is the highest occurrence of dengue fever all over the Pakistan. Lahore is the second largest city of Pakistan with an area of 1772 km^2 and among the most densely populated cities of the world with a population of over 8 million. It is a rapidly growing and developing city and has become Centre of attraction for the people willing to migrate from rural areas for better livelihood. This has caused a rapid increase in the population of the city during past years.

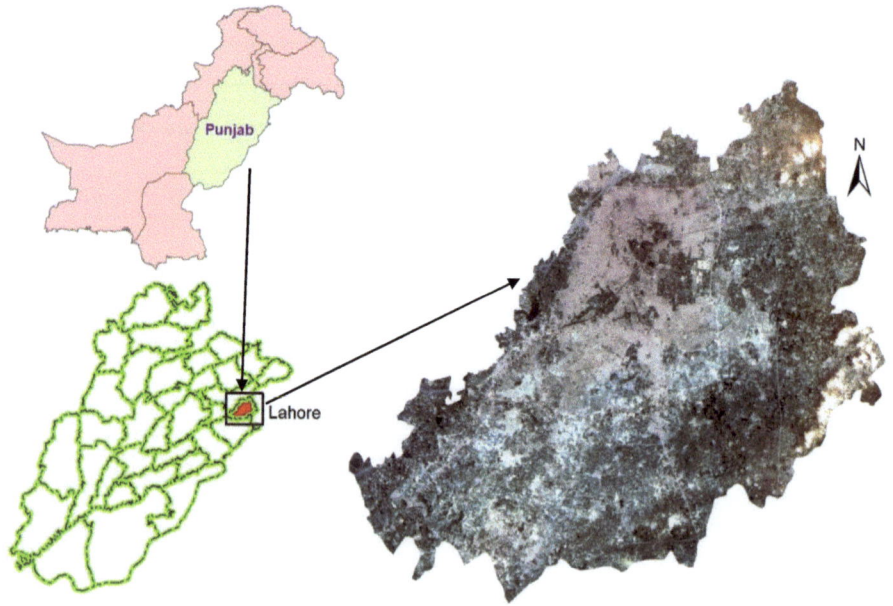

Figure 3.1: Geographic Location of Study Area in the Country

Study Area   Chapter 3

The Land cover/ Land use of the study area (built up, agricultural, swimming pools, stagnant water ponds, more area covered by parks etc.) in combination with the lack of strategic designs for mosquito sprays is observed to be quite suitable for the breeding of the mosquitoes.

According to the data collected by WHO (World Health Organization), more than 7000 people were victimized by dengue during the year 2011 only. The situation can become even worse if suitable actions would not been taken by the concerned authorities.

3.2 Geographic Location of Study Area

It is located in the geographic extents of 31° 15' and 31° 45' latitude and 74° 01' and 74° 39' longitude. The whole district is divided into nine towns which are further distributed in 150 Union Councils. A descriptive map of the towns and Union Councils of the study area is illustrated in the figure:

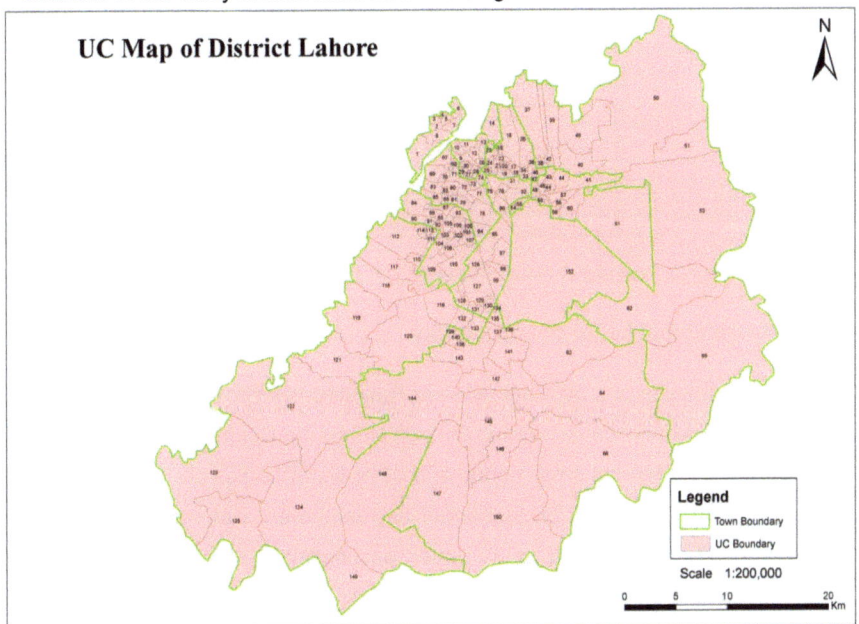

Figure 3.2: Map of Lahore District with its UC Boundaries

3.3 Land Cover of Study Area

Lahore is a metropolitan with a rapidly increasing population. The built up area of the city is increasing with each passing day. The imagery of Landsat V was classified using supervised classification technique with nearest neighbor interpolation and eventually got nine major Land Cover classed named; river, Canal, Vegetation, Low dense built up, high dense built up and medium dense built up, Soil, clouds and a mixed class where soil and low built up areas got mixed. The percentage of each type of land cover is shown in the table below:

Land Cover	Percentage (%)
River	0.44
Canal	4.75
Vegetation	41.52
Low Dense Built up	12.43
High Dense Built up	9.59
Medium Dense Built up	7.96
Soil	11.62
Mixed	9.42
clouds	2.26

Table 3.1: Land Cover Classification with its percentage

3.4 Hydrological Layer of Lahore City

Lahore is the city of a river names river Ravi passing through the eastern part of the city and a major canal passing through almost the Centre of the city. Moreover, there are a number of small ponds and other water bodies are also found in the metropolitan. They include permanent and temporary water ponds. Temporary water ponds include the small water accumulation sites where water gets accumulated for a number of reasons. They include rainfall and poor sewerage system etc. a detailed hydrological map of Lahore city is figured below:

Study Area   Chapter 3

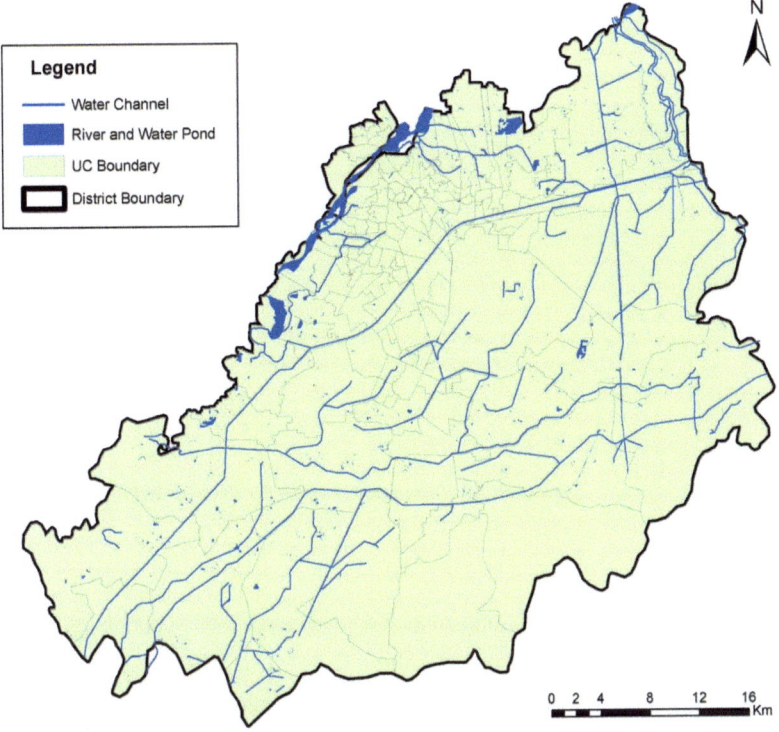

Figure 3.3: Map of Hydrological Layers of Lahore District

3.5 Climatic Conditions

Lahore lies in semi-arid climatic zone where the winters are quite dry and cold, summers are having extreme temperatures and a dry climate followed by heavy rainfall during the monsoon. Dust storms and thunderstorms also persist in the city.

3.5.1 Temperature

Lahore enjoys a wide belt of temperature range, i.e. starting from extremely cold climatic condition during the months of December and January reaching an extremely hot wave of summers in May and June. July, August and Mid-September are the months of monsoon. A clearer picture of the temperature profile of the year 2011 is represented graphically as:

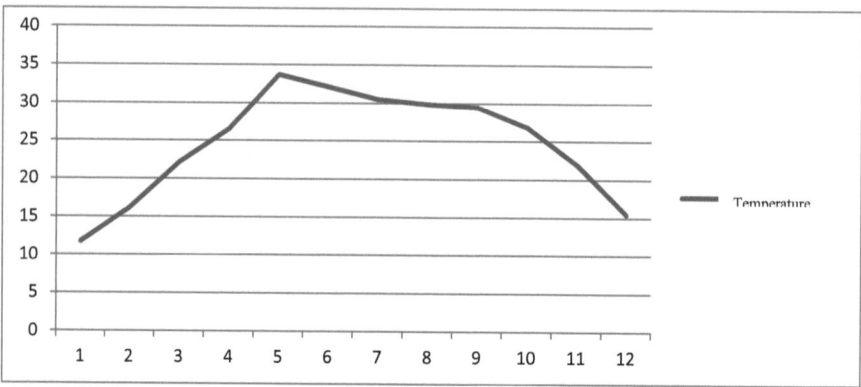

Graph 3.1: Temperature curve for the year 2011

3.5.2 Rainfall

Since Lahore lies in a semi-arid zone, it enjoys heavy rainfall during the months of monsoon. The average rainfall of Lahore is 470.1 mm (PMD). It is sufficient for good vegetation index as well as to fill the water bodies and water channels in the city. These water accumulation sites then become great breeding sites of the mosquitoes. This is the major reason behind the drastic increase in the number of patients during the months of monsoon. The graphical representation of rainfall of year 2011 and 2012 is given below:

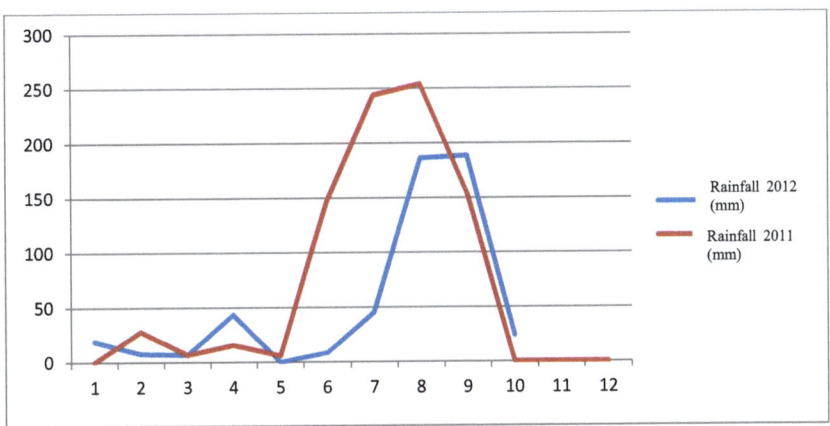

Graph 3.2: Comparison of rainfall in mm of the years 2011 & 2012

3.6 Suitable Breeding Sites for Dengue

The climatic condition and the land cover both are quite suitable for the breeding of Aedes mosquito. A good percentage of vegetation indirectly refers to the large number of water accumulation area due to watering the plants. They normally are small water accumulators but still are enough sized to breed mosquitoes. Secondly our construction pattern of most low income areas may

Dengue persists in mild temperature and humid climatic condition. The country's monsoon is very suitable for the breeding of mosquitoes with heavy rainfall and mild temperature. These are the main reasons behind the city being victimized by the dengue to such a large extent.

Chapter 4

Materials and Methods

4.1 Data Collection

4.1.1 Epidemiological Data

The first step involves data collection of the dengue patients for the year 2011. It was the most difficult task to do as, being a developing country and newly introduced to epidemic like dengue, there wasn't any proper collection of data in the previous year. The proper collection involves the no. of patients of dengue incidence, their addresses, gender, serotype, age, status, duration of admission in the hospital etc. We got the data from World Health Organization (WHO) having a few fields, i.e. Name of patient, age, gender and date of being admitted in the hospital along with the name of the hospital. Later on, this limitation resulted in a major hurdle in deducing quality results for our study.

Now, for the year 2012, different Govt. and private departments are very active to collect accurate data for dengue incidences to identify the spread pattern as precisely as possible to avoid losses of lives to a minimum number. A temporal analysis among the no. of confirmed dengue cases in 2011 and 2012 may help to figure out the accuracy of the available data of 2011.

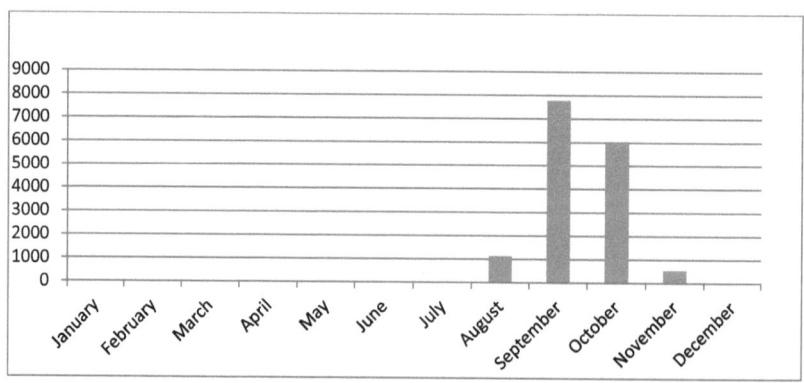

Graph 4.1: Graphical representation of Patient count for 2011

Month	No. of Patients
January	0
February	0
March	4
April	5
May	10
June	2
July	7
August	1123
September	7775
October	5987
November	541
December	0

Table 4.1: Monthly no. of reported patients of Dengue

4.1.2 Population Density

Another type of ancillary data which can be rated as secondary data than can help us in mapping the patterns of dengue incidence and find out the causes of such a spread pattern is population density data. The population density data was obtained from Punjab Development Statistics (PDS). This excel data was then brought into GIS environment and mapped accordingly.

Materials and Methods   Chapter 4

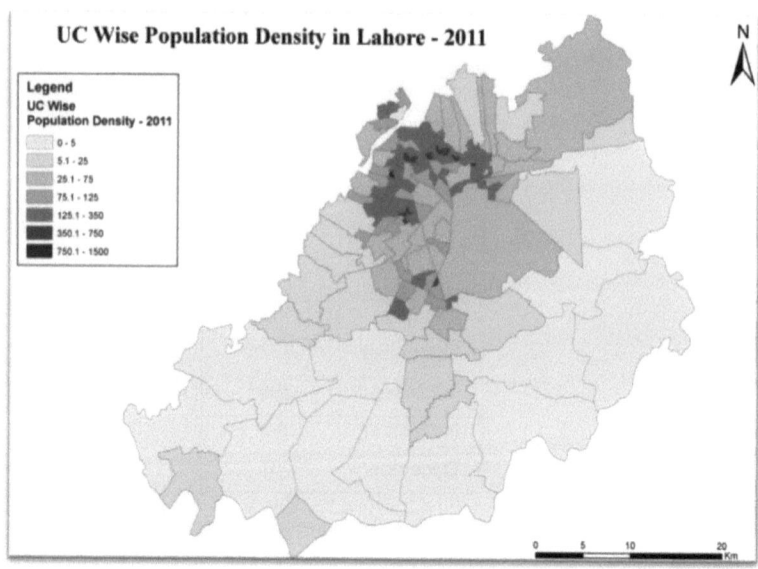

Figure 4.1: UC Wise population density of Lahore District - 2011

4.1.3 Identification of Environmental Parameters

The most important step of the research is the identification of the environmental parameters responsible for the existence and spread of dengue. A part of the meteorological data was extracted from remote sensing data. This methodology was adopted here because it was successfully implemented by a number of researchers. We followed the one tested by Connor et al., (1996).

The three major environmental parameters extracted from satellite imagery are Land cover, Normalized Difference Vegetation Index (NDVI) and Land Surface Temperature (LST) for the year 2011 whereas for 2012, this data was also obtained from PMD as Landsat V was no more operational in 2012.

The rainfall data was taken from the PMD (Pakistan Meteorological Department) for the year 2011.

4.1.3.1 Precipitation

The precipitation is any product of the condensation of the atmospheric water vapor that falls under the gravity. Among all the types of precipitation, rain has been selected as it causes water accumulation in the form of small stagnant ponds and other stagnant water deposits in the streets, roads, parks etc. It is observed that the major threat of dengue attack occurs in monsoon season when temperature is moderate and there would be enough rainfall for the breeding of dengue mosquito. The precipitation value for the year 2011 was found to be 856.9 mm and for 2012 was 531.5 mm (till October, 2012) (PMD).

4.1.4 Data from Landsat V

The following data has been extracted from Landsat V for September 2011.

- Land cover
- Land Surface Temperature (LST)
- NDVI

4.2 Methodology

4.2.1 Land cover/ Land use

Landsat-TM was used to derive land cover map for September 2011 in ERDAS Imagine 9.1. A total of seven classes were found including High built-up, Medium built-up, low built-up, soil, Mixed (Bare land & Soil), Vegetation and Water.

Materials and Methods   Chapter 4

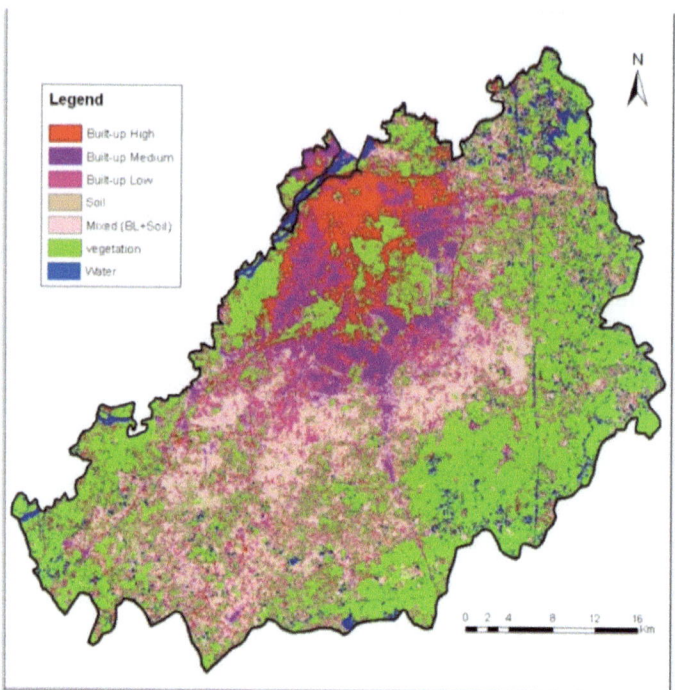

Figure 4.2: Land cover Classification done in ERDAS Imagine and mapped in ArcGIS 9.3

4.2.2 Land Surface Temperature (LST)

The LST was derived from the same Landsat-TM using band 6, named as thermal band. The Mono-Window algorithm was used to evaluate the LST using the surface radiance value from the thermal band. The algorithm adopted is mentioned as:

i: Conversion of the Digital Number (DN) to Spectral Radiance (L)

L = LMIN+ (LMAX-LMIN) * DN/ 255

Where

L = Satellite Spectral radiance

LMIN= 1.238 (Spectral radiance of DN value 1)

LMAX= 15.600 (Spectral radiance of DN value 255)

DN= Digital Number

Group scaling parameters' which shows the upper/lower bounds for radiance are in each band. We changed LMIN and LMAX for each thermal scene. (with the help of satellite header file.)

ii: Conversion of Spectral Radiance to Temperature in Kelvin

$$T_B = \frac{K_2}{\ln\left(\frac{K_1}{L_\lambda} + 1\right)}$$

Where

K_1 = Calibration Constant 1 (607.76) for LS-5

K_2 = Calibration Constant 2 (1260.56)

T_B = Surface Temperature

L_λ is the spectral radiance in W. (NASA, 2003.)

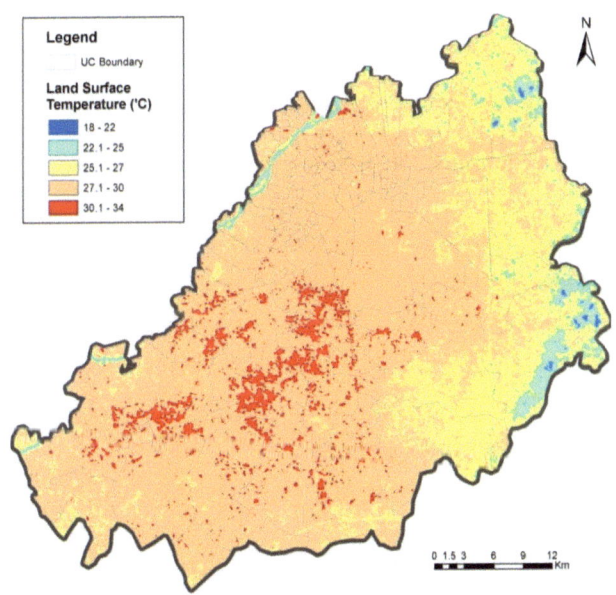

Figure 4.3: Land Surface Temperature for September, 2011

Materials and Methods   Chapter 4

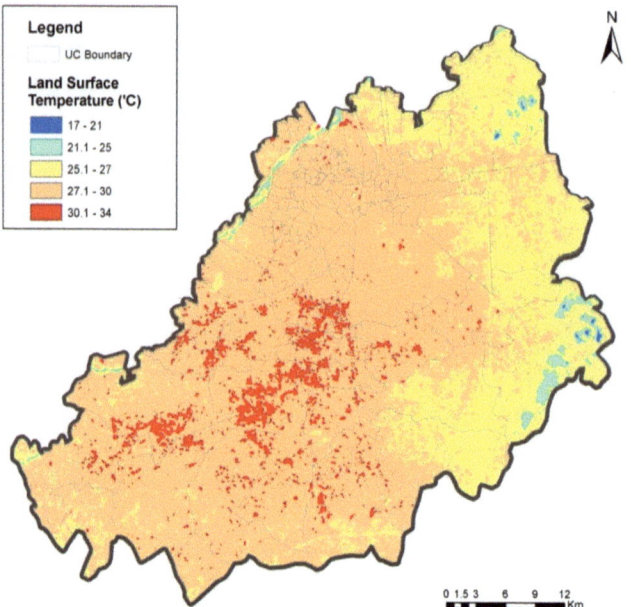

Figure 4.4: Land Surface Temperature for September, 2012

4.2.3 Normalized Difference Vegetation Index (NDVI)

NDVI (Normalized Difference Vegetation Index) is an important index to characterize the vegetation health and growth. (Hangbin et al. 2011). The NDVI has been calculated by using the visible and near infrared bands. NDVI indicated the vegetation health as well as differentiate between vegetated and non-vegetated areas. The formula used to find the NDVI of the satellite imagery of Landsat V of Lahore district is:

$$NDVI = (NIR - R)/(NIR + R)$$

It ranges between -1 to +1 for Landsat TM. The NDVI map for the study area is illustrated below with its upper and lower ranges in case of the current scenario.

Materials and Methods																																Chapter 4

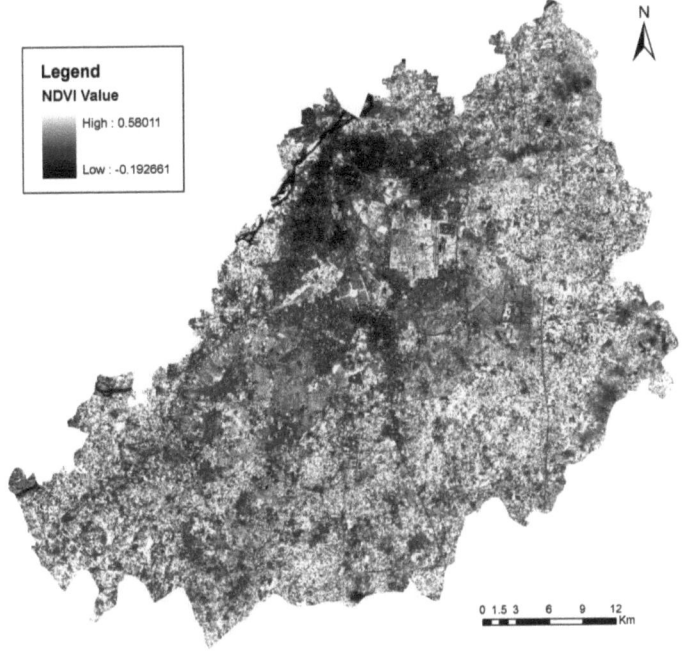

Figure 4.5: Normalized Difference Vegetation Index (NDVI)

4.2.4 Spatial Units to analyze and map Dengue Incidence data

The Union Council (UC) has been taken as the Spatial Mapping Unit for this study. Lahore is divided into 150 Union Councils. It covers an area of approximately 1772 km² with a population of 80,04,719 as per estimated in 2011 by PDS. The data collected from WHO clinical data was having the details of the patient's personal profile and the status of the fever. The addresses of the patients were spatially located (using geocoding technique) in the data for further analysis of the spread pattern of the disease.

4.2.5 Temporal Units

From the clinical data, it has been observed that there is a huge difference between the numbers of patients got caught by dengue in 2011 and the number of patients in 2012. The common factor was, the most threatening month in both years remained September. So a temporal analysis among the environmental parameters and the epidemiological data has been performed to figure out whether the pre-planning against dengue and the prevention measures taken by the city and provincial Government as well as the citizen gave some positive results in the form of lesser number of patients or not.

4.3 Data Analysis

All the data was brought to GIS environment and prepared accordingly. The above mentioned six variables (LST, NDVI, rainfall, Land cover, population and number of dengue patients) were then analyzed using weighted overlay technique. The weighted overlay tool facilitates us with the ease to combine multiple layers/ inputs to perform an integrated analysis by visualizing a map having the combine impact of all the selected parameters. The selected parameters were ranked according to the priority levels or effectiveness levels. The ranks are: low risk, medium risk, high risk and extreme risk for (1, 2, 3, and 4). This overlay analysis did not incorporate land use/ land cover classification as a separate layer to overlay because, the extraction/ estimation of land cover and land use is based on the spectral signatures. It was impossible to visually interpret the LC with a satellite imagery having 60m Spatial Resolution and no physical survey has been conducted till now for Lahore district. An estimated result about the most targeted Land Cover type can be interpreted in the light of the visual comparison of the risk areas and the Land Cover of the study area

Materials and Methods   Chapter 4

4.4 Identification of risk areas

The identification of dengue high risk zones/ areas was based on the analysis of the available medical data and the data extracted and prepared from the satellite imagery. The clinical data was brought into GIS environment to get a better visual identification of the areas to be focused. After identification of the most threatened zones, a raster map has been generated from the available data based on some preset criteria. It was analyzed that there is a direct relation between dengue incidences, rainfall, and surface temperature, high built up areas with a good NDVI value. (Nakhapakorn, Tripathi, 2005). The reason for each of the factor has been discussed in detail in the later sections.

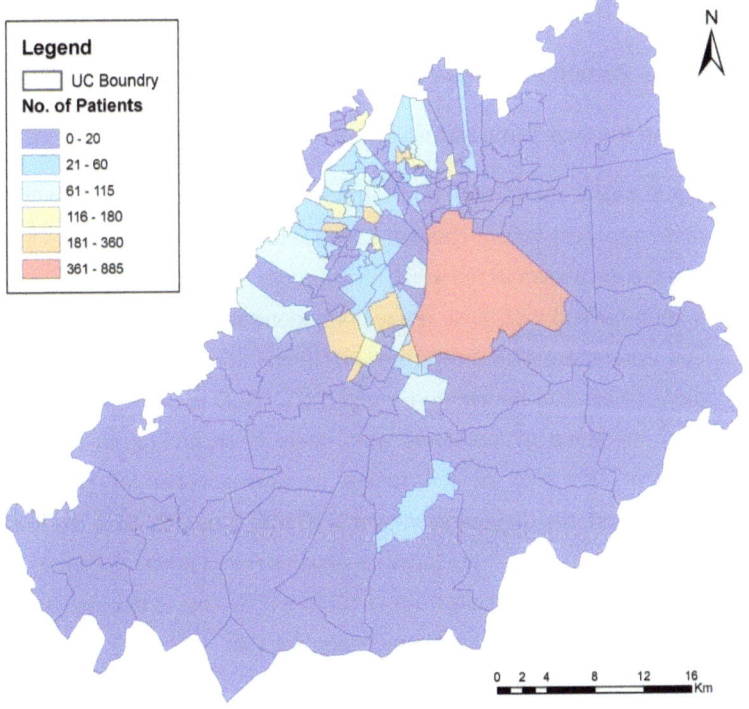

Figure 4.6: Map showing the most critical areas for dengue spread

Chapter 5

Results and Discussions

5.1 An Overview

The study started carrying a vague idea and some knowledge of literature about the dengue incidences. The key factors and the environmental parameters were selected after reading enough literature and a lot of research studies and they were tested to find out how and to what extent they influence the occurrence of dengue. With the data analysis, a number of expected results were found to be the facts which include: the relation of temperature and dengue incidences, the relation of dengue incidences with rainfall and population density etc. all the data was brought in the GIS environment and analysis, mapping and modeling was performed.

5.2 Spatial Pattern of Dengue Occurrences

5.2.1 Land Surface Temperature and Dengue Incidences

The figure below shows the Land Surface Temperature map obtained from the satellite imagery of Landsat Thermal Band TM (Band 6). The areas with the highest temperature are shown as red which are normally referred to as urban areas. Here, it has been noticed that these areas correspond to medium built up, Bare Land and Soil. Whereas, the areas with the least temperature are either vegetated areas or water bodies. The temperature range from 18° C to 34 ° C. Dengue is active mostly at moderate temperatures when there is an enough of humidity in the atmosphere. This kind of atmosphere is quite suitable for the breeding of mosquitoes and, the dengue incidence is directly proportional to the breeding of the mosquitoes. In actual, the larval development into adults occurs at moderate temperatures ranges typically from 16° C to 34° C. This is the reason why there occur a sever outbreak during the monsoon season.

Results and Discussions _____ Chapter 5

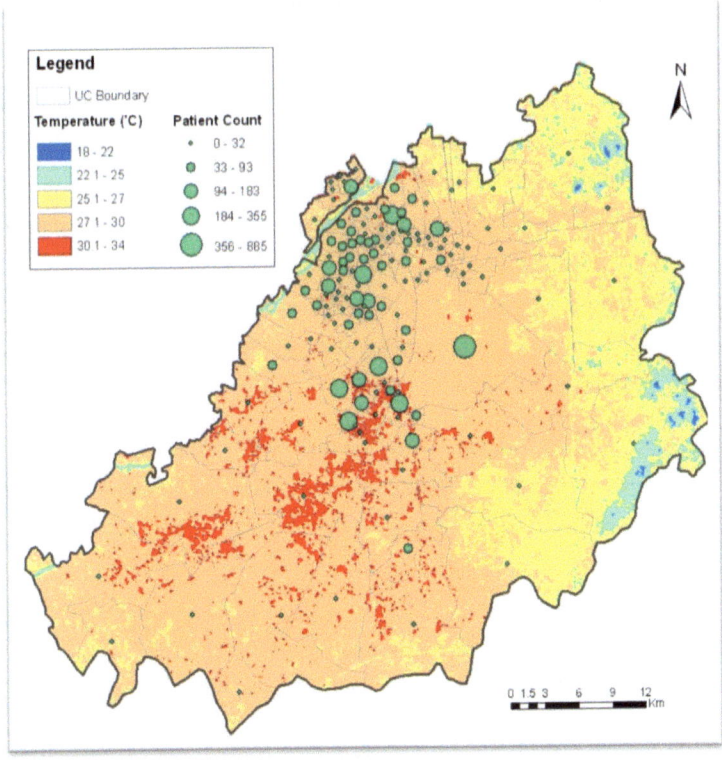

Figure 5.1: LST of 2011 with No. of patients in all UCs of Lahore

5.2.2 Land Cover Map against Dengue Occurrences

A Land Cover Map has been generated from Landsat (TM). It has a moderate spatial resolution of 60m. The image was classified into seven major classes, namely, High, moderate and Low built-up, Soil, Mixed i.e. Soil & Bare Land (BL) Vegetation and Water. From the classification analysis of Dengue patients against their number and spatial locations of the dengue patients, it is evident that the dengue cases were mostly occurring in the high built-up areas. The reason is the greater population density. The fact behind this kind of pattern is, the small radius of the activity area of Aedes Agypti. The Aedes travels with in an area of 320m (maximum limit).

From the map 5.2, it is also evident that the zones with the maximum dengue incidences were posh ones such as Cantonment, Johar town and Model Town having mostly pools and gardens (lawns) in their houses where there exist stagnant, clean water pools. A second and very important factor to be discussed here is about the acquisition of data. This data is acquired from World Health Organization (WHO) which comprises the collection of the data about the reported cases in the hospitals of Lahore. Here, the most important point is, the data is about the reported cases and it does not include the patients which have not consulted any hospital and relied on self-medication. So, the actual number of dengue incidences may differ.

This aspect may also be related with the land cover, as, is has been observed that the residents of the posh and high profile areas have a comparatively easy access to medical facilities. It might be a reason for the greater number of patients in posh areas like Cantonment, Model Town and Johar Town. On the contrary, some areas where people have an average income and the residents belong to middle class like Mozang, Ismail Nagar and Shadbagh are also found to have a comparatively greater number of patients. It can be justified by the fact, that, the radius of dengue mosquito is quite short i.e. only 320 m. (C.Liew, C.F.Curtis, 2004). These are congested areas where most of the houses are small in size. So, even a small number of mosquitoes is enough to infect a greater number of people.

Results and Discussions   Chapter 5

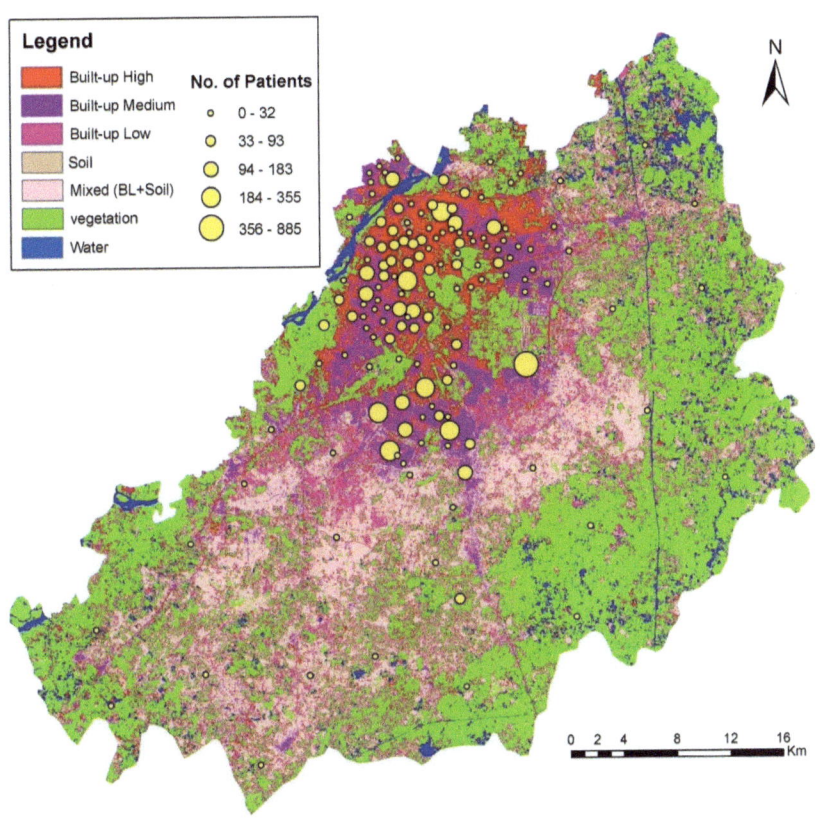

Figure 5.2: Land Cover Categorization of Landsat V image for Sept, 2011

5.2.3 Normalized Difference Vegetation Index

NDVI (Normalized Difference Vegetation Index) is used to differentiate between vegetated and non-vegetated areas. An NDVI map has been generated from Landsat TM satellite imagery. The built-up area is shown in dark patches whereas the vegetation is depicted in bright patches. It has been observed that there is no direct relation of NDVI with the number of patients. The only aspect which can be related to the number of patients is, the parks, gardens, lawns may have small pools of stagnant water which are favorable for the breeding of mosquitoes.

Secondly, when temperature is high, the water accumulated in these pools can be evaporated and the atmosphere around becomes more humid causing a rapid growth in the dengue breeding. This factor can be covered in Land cover mapping. The NDVI map is shown in the figure. Though this factor is not of much importance still it should be incorporated in mapping and can't be ignored.

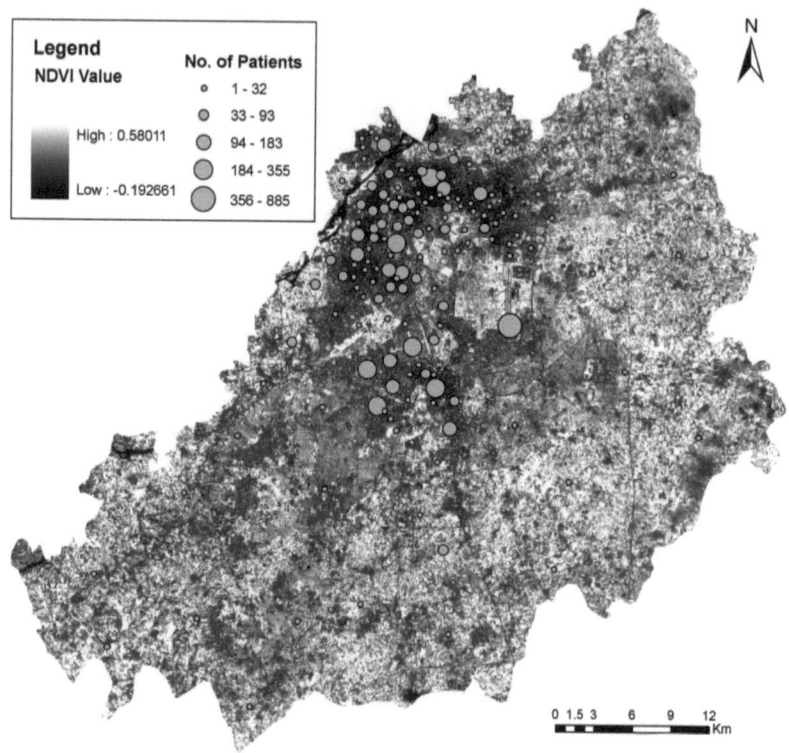

Figure 5.3: *NDVI in Comparison with the number of patients for Sept, 2011*

5.2.4 Population Density

This data also verifies the fact about the small radius of activity area of dengue mosquito. Densely populated areas indirectly refer to small size of houses. This factor again indirectly depicts the low income of the residents. So, the densely populated areas with a greater number of patients justifies the fact that the mosquito in that area gets a higher chance of finding a target and needs not to travel very far.

Another positive aspect of the greater figures of patient count is, these areas might have an easy access to the medical facilities and the awareness against the cure of dengue is greater in these areas (as we have mentioned earlier that this mapping and analysis is performed on the basis of reported cases). The analysis is also helpful for the Government to not only to find out the areas with high dengue incidence so that spraying can be done effectively in those zones, but, it also helps them to suspect why high population density areas and slums have low number of patients. Though it is good to have low patient count in such areas but still it is important to verify whether these areas actually have lesser number of patients or they haven't reported to the hospitals or medical camps and rely on self-medication due to financial issues or lack of awareness.

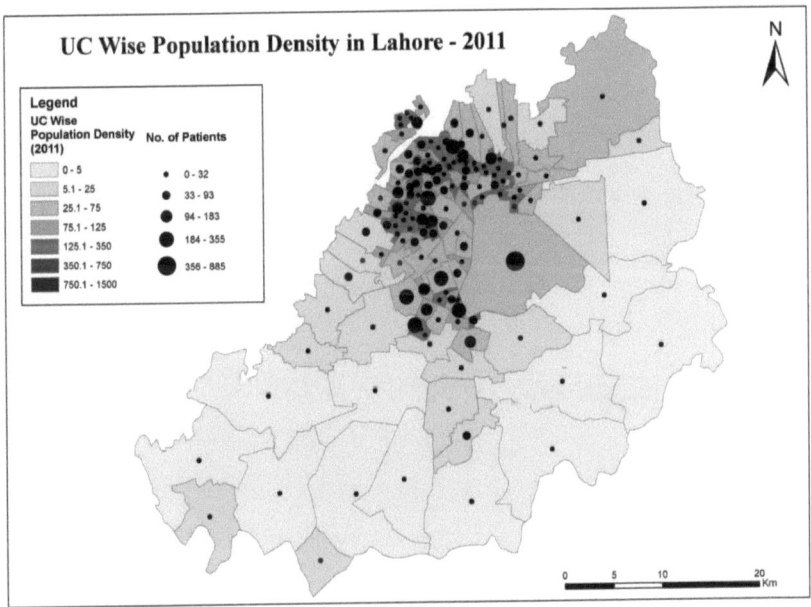

Figure 5.4: Population Density compared with the No. of Patients

5.3 Identification of High Risk Areas Using overlay Analysis for 2011

All the data of the reported dengue cases of Lahore District have been mapped UC (Union Council) wise demonstrating only the number of patients in each UC. A second map has also been generated overlaying all the ancillary data obtained from different sources.

The identification of high risk areas having a greater potential of being infected was determined using all the above stated parameters. The analysis was performed using the weighted overlay function by categorizing the factors according to their effectiveness on the dengue incidences. The ranks were, "low impact factor", "moderate impact factor", "high impact factor" and "very high impact factor". They were given the priority values 1, 2, 3 and 4 respectively.

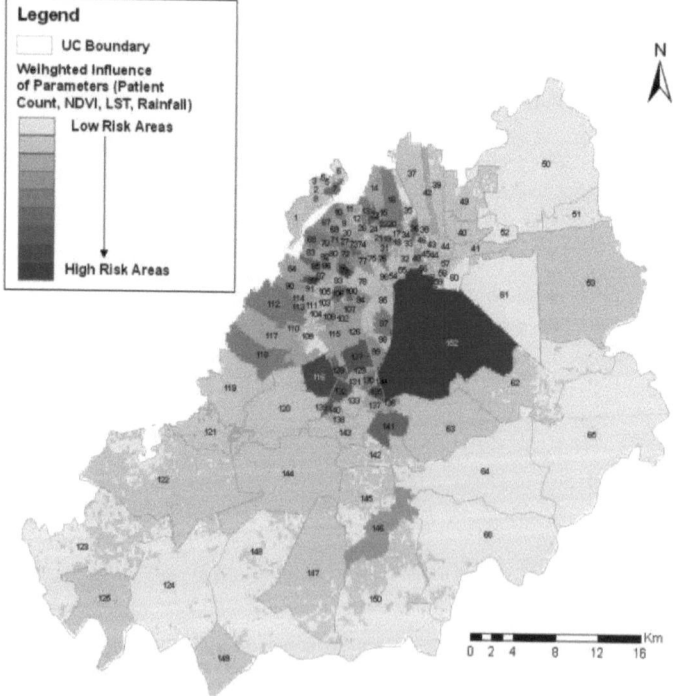

Figure 5.5: Accumulative map for spread pattern of dengue with all its causal parameters

Using this method, the areas with the highest risk are highlighted and mapped. The information depicted in both the maps can be correlated with each other showing the causes of greater outbreak in some areas in terms of climatic conditions and the extent of dengue outbreak.

Another important thing to mention is, there has been a different trend observed in terms of population density. So, the analysis including population density with all other parameters is kept separate. The results a highlighted few more zones as medium to low risk areas. We may conclude from the map that, our data for dengue patients was clinical data based on the cases reported in different medical centers. But being thickly populated zones, there might be some dengue cases which were not being reported. So, which performing the prevention activities, these areas should also be kept under consideration.

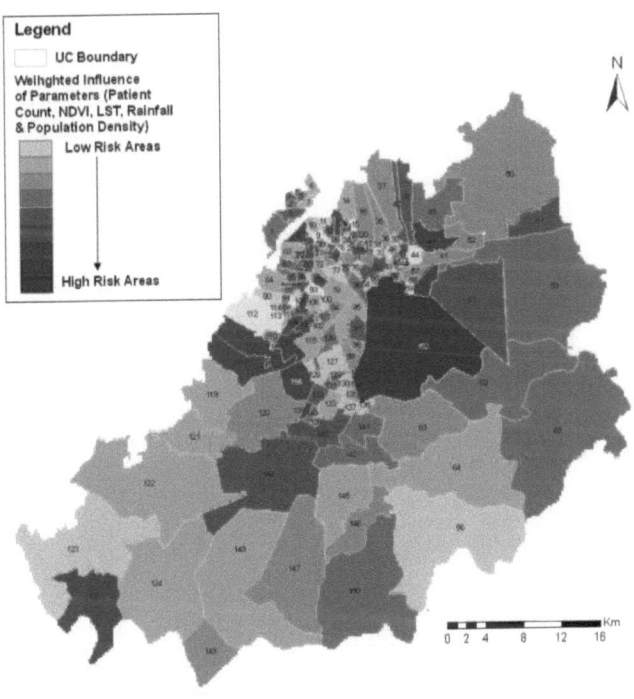

Figure 5.6: Accumulative map of dengue Spatial spread pattern with all the parameters

5.4 Temporal Analysis between 2011 and 2012

A temporal analysis has also been performed on the clinical and environmental data of the years 2011 and 2012. From the clinical data, it has been observed that there is a surprisingly rapid decrease in the number of patients for the peak month (September) of both 2012 as compared to 2011. In 2011, a total number of 7775 patients were reported in September whereas in 2012, this figure decreased to 2408. The reason behind this fact was investigated by comparing the environmental and epidemiological data of both the years. By this comparative analysis, it has been concluded that the control activities yielded great results reducing the total number of patients from 7775 to 2408 within a year. The reason can be well justified by observing all the parameters individually. There was no sudden change in temperature observed for these consecutive years. A decrease of 0.4 degree only cannot prove to be so affective that reduces a total of 7000 victims. Similarly, in a developing city like Lahore, there cannot be a drastic change in vegetation index and Land Cover types in just a year. The rain fall also remained average throughout the year. So, managed control activities resulted positively saving a lot of lives.

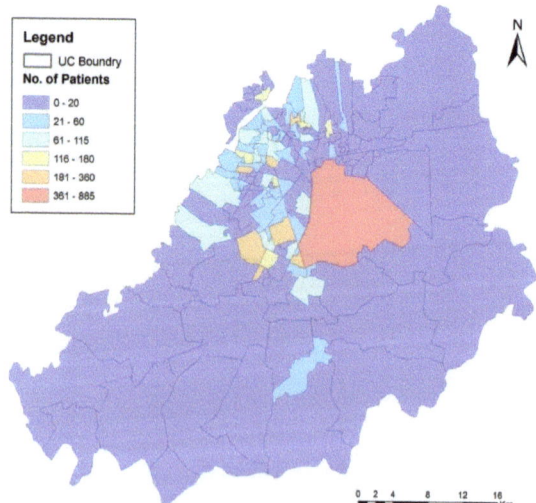

Figure 5.7: Dengue Spread pattern for the year 2011

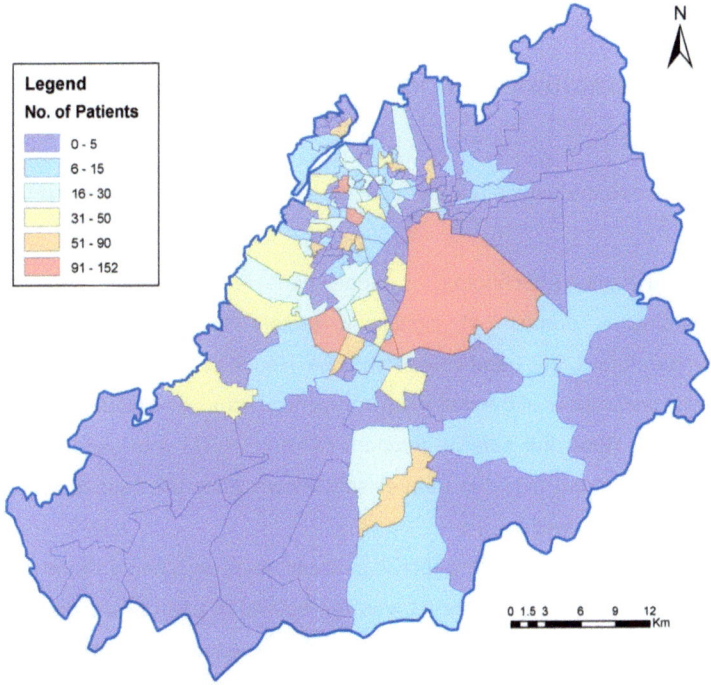

Figure 5.8: Dengue Spread pattern for the year 2012

Chapter 6

Conclusions and Recommendations

6.1 Conclusions

From all the above discussion and the analysis performed, the following facts can be concluded:

- Remote sensing technology is highly capable of providing information regarding a number of significant factors such as Land Surface Temperature (LST), Land cover/Land use, Health of Vegetation (NDVI) etc. These were some of the major environmental factors that influence dengue outbreak. Here, the satellite imagery of Landsat TM was used to obtain these factors but they can be precisely known using the data of IKONOS, ASTER etc. as well.
- The study depicted different spread patterns, each depicting a sound cause. One of the significant patterns was the spread in posh areas due to the reason that this mosquito breeds over clean water. The areas in these houses mostly have water pools and gardens in them which are suitable sites for the breeding of mosquitoes. Secondly this study is based on the number of reported cases, so being posh areas; the residents have an easy access to the medical facilities to report against the disease.
- Another pattern identified was a high number of patients in low profile areas having residents belonging to a middle class family. In this scenario, three major reasons were identified other than the environmental factors which were, a greater population density, least access to the medical facilities and lack of cleanliness. The unpaved roads causes the rain water to accumulate and form little water ponds which is the favorite site for the breeding of mosquitoes.
- The Remote Sensing in combination with the GIS technology has found to be very effective to plan against some disease surveillance system. The risk map has been generated taking all the environmental data obtained from

Conclusions and Recommendations   Chapter 6

different sources in the GIS environment using weighted overlay function. These are the initial findings which were made using the available data that depends on the precision of the data.

- The temporal comparative analysis showed that the control and prevention strategies have proved to be beneficial to a large extent. Continuing the same strategy with the same spirit may nullify this viral infection.

6.2 Limitations

There were certain limitations that restrict us to extract some more valuable results. These limitations are listed as followed.

- No data regarding the confirmed dengue cases was available. So we have to restrict out study over the suspected cases. There are chances that these suspected figures are not the actual ones as among 2286suspected cases of 2012, only 92 cases were the confirmed ones.
- The Landsat V stopped working in November 2011 (USGS Newsroom, 2011). So we were unable to extract the temperature for the coming year. So we used air temperature estimations to get the surface temperature of September, 2012. So the actual temperature may differ from this one.
- There must be a systemized clinical data of the serotypes of dengue to analyze the type of dengue virus dominating the others.
- Proper land use surveys also help to target the breeding sites of the mosquitoes so that all the control activities may focus those areas more actively.

6.3 Future Recommendations

There may certainly be more types of analysis that could be performed on during research but the limitations of data restricted us from doing so. A comparison of detailed surveyed land use data with the clinical data may be quite fruitful in getting the targeted zones of dengue in the form of small patches. Moreover, a temporal analysis of five or more years may also help to analyze and monitor the spread pattern of the disease more keenly.

References

Barbazan P, Amrehn J, Dilokwanich S, Gonzalez J-P, Nakhapakorn K, Oneda K, Thanomsinra A, Yoksan S: Dengue Haemorrhagic Fever (DHF) in the Central Plain of Thailand. Remote sensing and GIS to identify factors and indicators related to dengue transmission

Tzai-Hung Wen, Neal H. Lin, Chun-Hung Lin, Chwan-Chuen King, Ming-Daw Su. (2006): Spatial Mapping of Temporal Risk Characteristics to improve Environmental Health Risk Identification: A case study of a Dengue Epidemic in Taiwan

SIC (Satellite Imaging Corporation)

USGS (United States Geological Survey)

Gong Peng X B (2006). Remote sensing and geographic information systems in the spatial temporal dynamics modeling of infectious diseases. Science in China Series C: Life Sciences, 49, 573-582.

Haja Andrianasolo S Y P (2001). Remote sensing in unravelling complex associations between physical environment and spatial classes of emerging viral disease. 22nd Asian Conference on Remote Sensing. Singapore, 5-9 November 2001.

Krishna Prasad Bandari P R (2008). Application of GIS modelling for dengue fever prone areas, based on sociocultural and environmental factors – A case study of Delhi City Zone. The International Archive of The Photogrammetry, Remote Sensing and Spatial Information Sciences, XXXVII, 165-170.

Maynard N G (2002). Remote sensing for the public health surveillance and response. Earth Observation Magazine. Special NASA Earth Science Enterprise Issue, pg 43-45 [accessed 28/09/09].

M.N. Bayoh, S.W. Lindsay, June 2, 2004: Temperature Related duration of aquatic stages of the afrotropical malaria vector mosquito anopheles gambiaein the laboratory,

Tzai-Hung Wen et. al (2002): Spatial–temporal patterns of dengue in areas at risk of dengue hemorrhagic fever in Kaohsiung, Taiwan,

Mondini et. Al (2008): Spatial correlation of incidence of dengue with socioeconomic,

demographic and environmental variables in a Brazilian city

Ratana Sithiprasasna K J (1997): Use of GIS to study the epidemiology of Dengue Haemorrhagic Fever in Thailand. Dengue Bulletin, 21, 68-73.

C. Liew, C. F. Curtis (2005): Horizontal and vertical dispersal of dengue vector mosquitoes, Aedes aegypti and Aedes albopictus, in Singapore